RESOURCES FOR PHYSICS 21040

Physics in Entertainment and the Arts

A COURSE IN FOUR ACTS

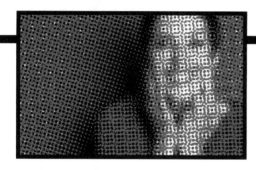

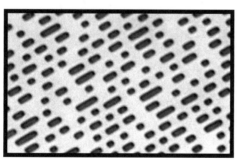

Jon Secaur

VAN-GRINER

Physics in Entertainment and the Arts

Resources for PHY 21040
A Course in Four Acts
Jon Secaur
Spring 2015

Cover images: Top bridge photo is a screenshot from https://youtu.be/j-zczJXSxnw. The bottom image is a microscopic view of the pits and smooth areas in an audio CD. Each pit is about 1/30th the width of a human hair. © Eye of Science/Science Source.

Photos and other illustrations are owned by Van-Griner or used under license.

Printed in the United States of America
10 9 8 7 6 5 4 3 2 1
ISBN: 978-1-61740-320-0

Van-Griner Publishing
Cincinnati, Ohio
www.van-griner.com

Secaur 320-0 Su15
Copyright © 2016

Some Resources for Act III

FRAZZ   **BY JEF MALLETT**

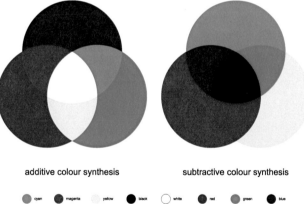

additive colour synthesis   subtractive colour synthesis

cyan   magenta   yellow   black   white   red   green   blue

What's wrong with this image from shutterstock.com, a professional graphic arts supplier? They should know better!

A Little Help with Waves

If these seem easy to you, please be patient with those who don't think so. These questions all refer to the American Association of Physics Teachers video on the Tacoma Narrows, Washington, bridge disaster in November, 1940.

Items with check marks (✓) are facts you must listen for in the narration.

✓ 1 | How long was the central span of the bridge?

2 | On normal days, what kind of wave (transverse, longitudinal, or torsional) moved through the bridge?

✓ 3 | What was the total up-and-down motion of the bridge on normal days?

4 | Following up on Question 3, what was the amplitude of the bridge's motion on normal days?

5 | Using the stopwatch on my phone, I measured an average of 7.8 seconds for three total up-and-down cycles of the bridge's motion. From that data, what was the period of the motion?

6 | What was the frequency of the motion?

7 | Use the still image from the video, at the right, and Question number 1 to estimate the wavelength of the wave that was moving through the bridge. Notice that between the two support towers, the edge of the bridge toward us moves up, then way down, and back up to its normal position.

https://youtu.be/j-zczJXSxnw

8 | Using answers from Questions 6 and 7, what was the speed of the wave along the bridge?

9 | On November 7, 1940—the day the bridge collapsed—what kind of wave (transverse, longitudinal, or torsional) moved through the bridge?

10 | From measurements on my computer screen and the fact that the side girders were eight feet high, I estimated that the amplitude of the waves that destroyed the bridge was about 15 feet. What was the total up-and-down motion of the bridge?

A Little Help with Waves, Take Two

1 | A stop sign twists back and forth during a windy blizzard, making three complete oscillations every second.

    a | What is the frequency of that motion?

    b | What is its period?

    c | If the edge of the sign moves back and forth a total of 6 cm, what is the amplitude of the motion?

    d | What kind of vibrational motional is that?

2 | If you are listening to a concert by your favorite band, what one thing must be the same for all of the sound waves that you hear?

3 | How would two sound waves sound to you, if they have …

    a | the same frequency but different amplitudes?

    b | the same amplitude but different frequency?

    c | the same amplitude and frequency, but different phase?

4 | Sound waves in air travel at about 340 m/s.

    a | What is the wavelength of a 440 Hz note?

    b | What is the frequency of the second harmonic of that note? Of the third harmonic?

5 | Two trumpet players are tuning up for a concert. One plays a note 320 Hz, and the other, at 324 Hz.

    a | Which one is sharp?

    b | What will they hear?

    c | What is time between beats called? How long would that be in this case?

6 | To make a second harmonic on a 2.0 m long rope, you have to move your hand up and down four times a second.

    a | What is the frequency of the second harmonic? What's the frequency of the fundamental or first harmonic?

    b | What's the wavelength of the second harmonic? Of the fundamental?

    c | How many times per second would you have to move your hand to make the third harmonic? The fourth?

7 | The village of Humdrum has a wishing well in the town square. They call it a wishing well because every time the wind blows, the well makes a deep 20 Hz humming sound and everybody wishes it would stop. Recall that a well is a pipe closed at one end. How deep is the well?

8 | A piece of downspout tied across the bed of a pickup truck has been humming at 55.7 Hz all the way home from the lumberyard. The driver pulls into his garage, which has a door that is 3.0 m wide. Will the downspout fit through the doorway? Assume that it was open at both ends (at least until he tried to pull in!)

"Numerology" in Harmonics on Strings

Other titles could have been, "Mathematical relationships among harmonics on strings," or "Naming and numbering harmonics on strings." The first sounds too difficult, and the second doesn't go far enough.

A standing wave is a wave that fits in a bounded medium, or as we saw previously, a whole set of waves that fit. The waves in that set of waves that fit are called **harmonics.** That's because if a wave is only, say one-half or one-third as long as one that fits, then two or three of those shorter waves, respectively, will also fit.

This drawing shows the first six members of a set of waves that fit in the space between the red lines.

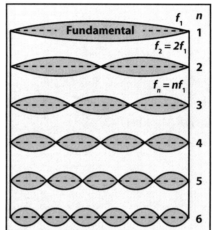

The grand-daddy of them all, the largest wave that can fit in the space, is called the first harmonic or the fundamental.

In this drawing its frequency is denoted by f_1 for the "frequency-of-the-first-harmonic," although some web sites and books designate it as f_o for "frequency-original." For the next wave to fit, its frequency must be twice as great, so $f_2 = 2 f_1$. (How do we know that? Since all the waves on in the same medium—the same string—they all travel at the same speed. The attributes of a wave are related by $v = f\lambda$ so if the wavelength, λ, needs to be only half as long, the frequency, f, must be twice as great to keep v the same.)

Then, the frequency of the third harmonic must equal three times the fundamental, and so on for all harmonics. In general, we can say that the frequency of the nth harmonic is n times the fundamental frequency, or

$$f_n = nf_1$$

What can we say about wavelength? You must remember that we **"catch" only half the wavelength of the fundamental.** That's because one wavelength is the length of a whole wave, from any one point on a wave to the same point on the next wave. If we trace the fundamental in the drawing, we start out from the left end at zero displacement, and going up. After tracing the entire drawing, we are back to zero displacement, but going down. We would have to move through that much of the drawing, again, to be moving up again at the zero displacement point.

All of that is to say, **the length of the string is exactly half the wavelength of the fundamental,** or, one-half wave fits in the fundamental mode. We just saw that the second harmonic is half the wavelength of the first or the fundamental, so **for the second harmonic—only—the length of the string equals the wavelength of the wave.**

Each antinode or segment or loop or "sausage link" in the wave is one-half wavelength. So, for the third harmonic, three half waves fit, and the string length is 3/2 or 1.5 wavelengths. If we consider the fourth harmonic, four half-waves fit, and the string length is 4/2 or 2 wavelengths.

Let L represent the length of the string. We can then summarize all of this in the following table:

HARMONICS			
HARMONIC NUMBER, n	NUMBER OF ANTINODES OR LOOPS	FREQUENCY OF THE HARMONIC	HOW MANY HALFWAVES FIT
1	1	f_1	$L = \frac{1}{2}\lambda$
2	2	$2f_1$	$L = \frac{2}{2}\lambda = \lambda$
3	3	$3f_1$	$L = \frac{3}{2}\lambda$
4	4	$4f_1$	$L = \frac{4}{2}\lambda = 2\lambda$
n	n	nf_1	$L = \frac{n}{2}\lambda$

As I like to say, **harmonics are your friends.** The number of loops or antinodes or "sausage links" you see in a drawing of a standing wave on a string tells you immediately the

- harmonic number,

- multiple of the fundamental frequency, and

- number of half-wavelengths that fit.

For example, what can we learn from this image?

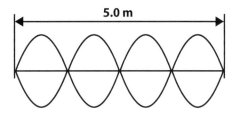

5.0 m

Since there are four segments or loops or sausage links, we know immediately that it would be called the fourth harmonic. Further, since we know that the string is 5.0 m long, each segment is 1.25 m. Each segment is also one-half wavelength long, so the wavelength is 2.5 m.

Suppose we also knew that the frequency of this wave is 40 Hz. The frequency of the fundamental would then be 40 Hz/4 = 10 Hz, and waves with frequencies of 20 Hz, 30 Hz, 50 Hz, and 60 Hz (and many more!) would also fit on that 5.0 m long string. But waves with frequencies of 5 Hz or 25 Hz or 31.4 Hz would not fit.

Decibel Scale Practice

Remember: an increase in intensity of 1 bel or 10 decibels (10 dB) means 10 times more intensity.

More formally, $\qquad \beta = 10 \text{ dB } [\log (I/I_o)]$

where β is the relative intensity of two sounds, in decibels, I is the greater of the two intensities you wish to compare, and I_o is the smaller. Both intensities are measured in W/m^2 or, sometimes, in W/cm^2. All that matters is that both I and I_o must be in the same units.

Also recall that the perceived loudness is proportional to the square root of the relative intensity, or **an increase in intensity of 2 bels or 20 decibels (20 dB) means approximately 10 times louder.**

Working backwards, we can find the relative intensity of two sounds—how much more intense one is than another—by

$$\text{relative intensity} = 10^{[(\beta_2 - \beta_1)/10]}$$

$$\text{relative loudness} = 10^{[(\beta_2 - \beta_1)/20]}$$

Example 1

What is the difference in intensity level, in dB, between 5.0×10^{-6} W/m^2 and 2.2×10^{-8} W/m^2?

$$\beta = 10 \text{ dB } [\log (5.0 \times 10^{-6} \text{ W/m}^2 / 2.2 \times 10^{-8} \text{ W/m}^2)]$$

$$= 10 \text{ dB} \times 2.36$$

$$= 23.6 \text{ Db}$$

Example 2

How much more intense is 53 dB than 50 dB?

$$\text{relative intensity} = 10^{[(53 - 50)/10]}$$

$$= 10^{(3/10)}$$

$$= 10^{(0.3)}$$

$$= 2$$

Any difference of 3 dB means that one sound is twice as intense as the other. A difference of 6 dB is approximately twice as loud.

For practice. The first six would be fine test questions.

1 | How much more intense is 75 dB than 35 dB? About how much louder would 75 dB sound?

2 | What is 1/10 as intense as 40 dB? What would sound 1/10 as loud as 40 dB?

3 | What factor covers our entire range of hearing, from the threshold of hearing at 0 dB to the threshold of pain, 120 dB?

4 | What intensity, in decibels, would seem approximately 100 times louder than 20 dB?

5 | Noise has been a concern for city dwellers since Roman times when rulers passed a bill that prohibited chariot driving through the cobblestone streets of Rome at night. Suppose a chariot rumbled near your Roman home producing a sound intensity of 80 dB. How much more intense is that than your spouse's whisper at 20 dB? About how much louder would it sound to you?

6 | Why do questions regarding loudness usually include words such as "approximately" or "about?"

7 | If a difference of 10 dB means 10 times the intensity, you might think that a difference of 5 dB means 5 times the intensity. Is that true?

8 | How much louder is 48 dB than 35 dB?

Harmonics in Human Speech

'OOH'

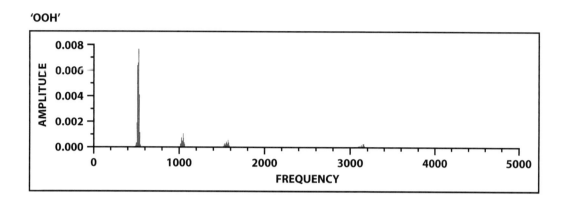

'AAH'

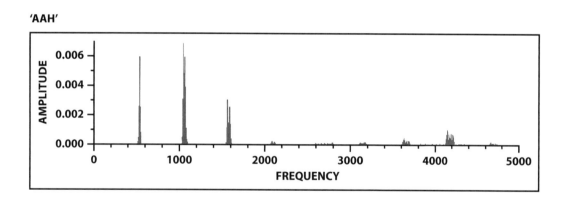

'EEH'

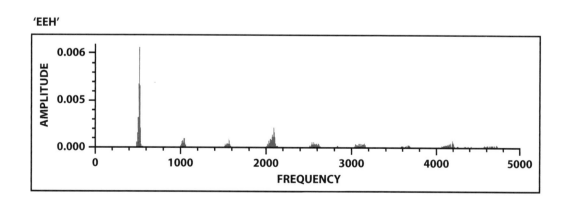

Harmonics in Musical Instruments

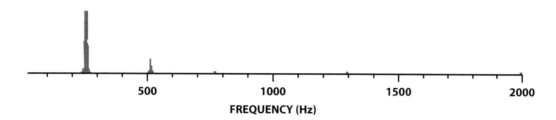

Here you see the Fourier graph or sound spectrum of a tuning fork playing a middle C. A perfect graph of a tuning fork would show one line only for the fundamental or first harmonic. That's almost what we have, but with a little of the second harmonic, as well.

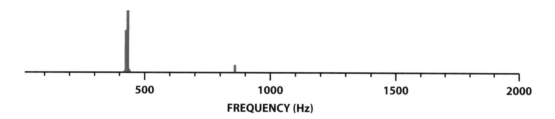

Sounding a shorter tuning fork produces a higher pitch, an A above middle C. The first harmonic shifts to the right, and the little stub of second harmonic moves as well. The whole graph stretches to the right, keeping equal spacing between the harmonics.

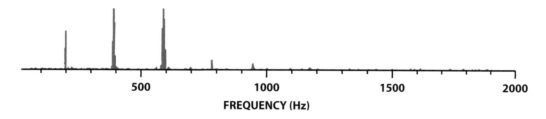

A mandolin produces a more complex sound, indicated by the busier sound spectrum. The first line from the left—the fundamental or first harmonic—still sets the pitch, even though it is not the strongest or tallest. All of the other harmonics, together, determine the quality of the sound, its "mandolin-ness."

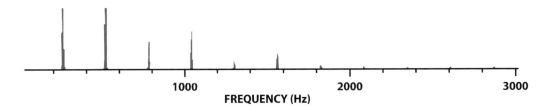

A reed instrument, such as this saxophone, creates this interesting and complex pattern.

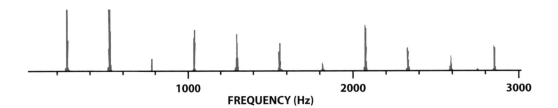

If that same saxophone is blown harder or louder, the first few harmonics stay about the same, but the higher frequency harmonics grow more, indicating a reedier, harsher sound. In general, when you sing or play more loudly, you add more high harmonics, making a more distinctive but often less pleasant sound.

Amazing Facts about Light

1 | Light is an *electromagnetic wave.* A changing electric field (such as the motion of an electron) makes a changing magnetic field, and that changing magnetic field makes a new, changing, electric field, which makes a new, changing, magnetic field, which makes a new electric field, which makes a new magnetic field, which makes a new electric field ...

2 | Light travels *fast!* Here's a table of how far light can travel in various times:

HOW FAR LIGHT CAN TRAVEL	
UNIT OF TIME	**DISTANCE LIGHT TRAVELS**
nanosecond (10^{-9} sec)	
microsecond (10^{-6} sec)	
millisecond (10^{-3} sec)	
second	
	from the sun to the earth
year (3.15×10^7 sec)	

3 | Light we see is only the tiniest slice of all the forms of light. If a chart of the entire spectrum from radio waves to gamma rays could be represented by a reel of movie film 2500 miles long, the part we see would be represented by _____.

4 | All we ever see is light! Objects have no color, and neither does light. Color is entirely _____. What we call color is a function of the _____ of a light wave.

Color Theory in Physics

Light does not have color, and objects don't, either. Color is a sensation as a response to frequencies of light, just as pitch is a sensation in response to frequencies of sound.

COLOR MIXING	
ADDITIVE	**SUBTRACTIVE**
For Light *Sources* Things that produce their own light (televisions, monitors, theater lights …)	**For Light *Suckers* or *'Sorbers*** Things that do not produce their own light (inks, paints, dyes, pigments …)
Three primary colors (all one syllable) **Red** – lowest frequencies we can see **Green** – middle frequencies **Blue** – highest frequencies we can see	**Three primary colors** (2 or 3 syllables) **Magenta** – absorbs *green* **Yellow** – absorbs *blue* **Cyan** – absorbs red (*shortest color words*)
Red and green give the sensation of yellow. Red and blue give the sensation of magenta. Green and blue give the sensation of cyan.	Magenta and yellow appear red under white light. Magenta and cyan appear blue under white light. Yellow and cyan appear green under white light.
All three give the sensation of white. To make black? No lights at all.	All three absorb all light and appear black. To make white? No pigments at all. Note that the traditional primary colors of red, yellow and blue are just renaming magenta, (yellow), and cyan.
For television production, stage lighting …	For visual arts, graphic arts, fashion design …

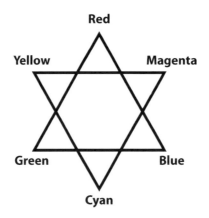

Color Addition and Subtraction Practice

1 | Television and computer monitors produce only _____, _____, and _____.

2 | Complete this table for colors in televisions and computer monitors by marking a check mark (✓) in the column if that color is present, and an ✗ if it is not, or by writing in the appropriate color word.

COLORS IN TELEVISIONS AND COMPUTER MONITORS			
RED	**GREEN**	**BLUE**	**WHAT WE SEE**
✓	✓	✓	white
			yellow
			black
✓	✗	✓	
✗	✓	✓	

3 | How would a television produce a pale blue, say, for a sky?
   HINT: Think of pale blue as a combination of blue and white.

4 | Color printers use three kinds of ink, that under white light look _____, _____, and _____.

5 | When any color of light falls on something white, you will see _____.

6 | When any color of light falls on something that is truly black, you will see _____.

7 | When any one primary color of light falls on an object of any color, the only choices for what you can see are _____ or _____.

8 | When any composite color of light falls on an object of any color, there are four possible outcomes: _____, _____, _____ or _____.

9 | Consider an American flag.

Complete this table for what colors you will see with each color of incoming light.

THE COLORS YOU WILL SEE			
INCOMING LIGHT	RED STRIPES	WHITE STRIPES AND STARS	BLUE FIELD
red			
green			
blue			
magenta			
yellow			
cyan			

How *All* Light Is Produced!

Step 1

In a normal atom in the **ground state,** electrons are as close to the nucleus as possible, and happy to be there.

Step 2

Energy is added—from an impact from another fast electron, or from heat, electricity, light, or a chemical reaction Note that this cartoon shows both the particle nature of the electron (the popcorn-looking blob bouncing off) and the wave nature (the more-or-less circular rings around the nucleus).

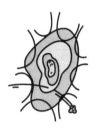

Step 3

An electron absorbs a quantum of energy, disappears from the real universe, and reappears at another, higher energy level. It's called a **quantum leap** because it is discontinuous—the electron doesn't slide out to a larger orbit, it just re-forms there! (Does it make any sense to say it's the same electron?) This is called the **excited state.**

Step 4

The excited electron disappears again, reforming at a lower energy level. Born with it is a **photon of light** that zips away. **All light in the universe is made this way!** The atom returns to the ground state, as in Step 1.

Usually the excited electron falls immediately, but there are three interesting variations. In **phosphorescence** or "glow in the dark" electrons can hang in Step 3 for seconds or even minutes before falling. **Fluorescence** or "Day-Glo" is when the input energy is supplied by invisible, high energy ultraviolet light and Step 4 is repeated several times, releasing lower-energy visible light. Lasers work by **stimulated emission,** where a passing photon triggers or tickles an excited atom in Step 4. The resulting new photon is exactly in step with the one that triggered it.

Blind Spot Test Targets

Using your right eye only, look at the ✕ (or, your left eye only, and look at the ✪).

Hold the page upright and level, about a foot away from your eyes. Move it in slowly, keeping your gaze on the symbol. You should see the other symbol disappear as its image falls on your eye's blind spot.

Want to Have Some More Fun?

Copy this page onto colored paper—any color of your choice—light, dark, bright, pastel …

When the symbol disappears, you won't see white, but the color of the paper. Your wonderful brain "samples" the background color and uses it to fill in over the blind spot!

This is one of the best confirmations that vision is an *active* system requiring lots of processing by our brains.

Making Holograms

Making Holograms

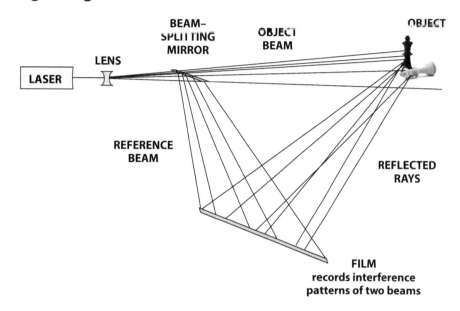

Viewing Holograms

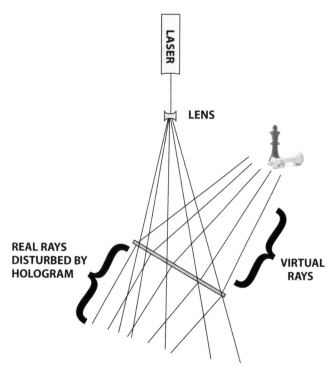

Finding Your Way Around Lenses

LENSES				
LENS TYPE AND ILLUSTRATION	**OBJECT DISTANCE**	**IMAGE DISTANCE**	**IMAGE TYPE**	**IMAGE SIZE**
 Concave or Diverging ☺	Application Example:			
 Convex or Converging	very large			
	Application Example:			
 Convex or Converging	> 2f			
	Application Example:			
 Convex or Converging	2f			
	Application Example:			
 Convex or Converging	between f and 2f			
	Application Example:			
 Convex or Converging	f			
	Application Example:			
 Convex or Converging	< f			
	Application Example:			

Help with Refraction and Lenses

1 | Here we have identical glass blocks suspended somehow in two different liquids. Which liquid has an index of refraction more nearly like that of glass?

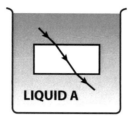

2 | A ray of light makes its way through three slabs of transparent solids, refracting at each surface, as shown. (Some light would also reflect from each surface, but we're only thinking about refraction, here.) Rank the three substances from the lowest index of refraction to the highest.

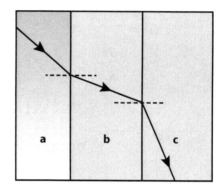

3 | Lenses for most 35 mm cameras have focal lengths of 50 mm or 55 mm. Suppose that a dealer in used cameras has an old lens with the paint worn off, and he's not sure which of those two lenses he has. He finds that the image of the store window, 5.0 m away, comes to a focus on a piece of paper held 55.6 mm behind the lens. Which kind of lens is it, 50 mm or 55 mm?

4 | If you want to start a campfire by focusing the image of the sun through a 50 mm lens, how far above the kindling should you hold the lens?

5 | A converging lens produces an image of a candle on a screen. The screen is 40 cm from the candle and the lens is positioned halfway between the candle and the screen. What is the focal length of the lens?

6 | If an object is placed 200 mm in front of a 100 mm focal length lens, where will the image be?

7 | What can you say about the shape of the lens in this illustration? (Besides *round.* ☺)

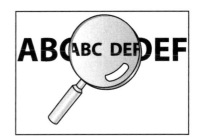

8 | There are two trees outside your window, one twice as far away as the other. You hold a white cardboard screen behind an apparently very large lens and bring the image of the nearer tree into focus. Which way would you move the screen to focus on the image of the farther tree— toward the lens, or away?

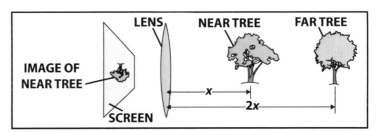

9 | Here's that image from the Question 2, again, but with the starting and ending points of the light's path marked. What can you say about that path from one **✕** to the other?

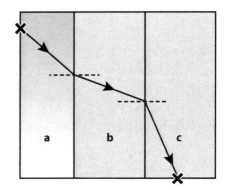

10 | Recently I visited the Fostoria Art Glass factory in Marietta, Ohio, while on a trip near there. Among many other things they make are large, spherical solid glass paper-weights with amazing colored glass designs inside. I thought of physics when I saw the sticker attached to each one that says "NO direct sunlight." Why must you not use this glass paper-weight in direct sunlight?

PHYSICS IN ENTERTAINMENT AND THE ARTS

Analog and Digital Information

Have you ever heard the old joke about how a teacher grades essay papers? The teacher stands at the top of the stairs and tosses the papers into the air. Papers that land on the first step get As, ones on the second step get Bs … Ha!

If you understand that, you understand the way **analog** signals are converted to **digital** signals! Analog information is continuously variable—for example, the position of a bug on the sloped floor of the classroom, or the amplitude of a wave at any instant. The term comes from the word *analogous* or *analogy,* because the wave that carries the music, for example, does whatever the original music did. But analog signals can easily be distorted along the way, so all modern media are digital instead, and we'll see why. The wave in the drawing below is an analog signal. At three points marked t_1, t_2, and t_3, the displacement or height of the wave is measured, or **sampled.** That height is "thrown against the stairs"— that is, the height is compared with a set of hundreds or thousands of reference values, and the step number of the closest one is recorded in **binary code.**

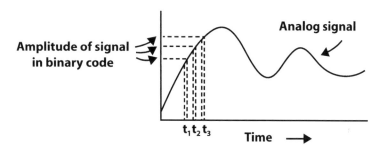

Let's try out some binary numbers! You know in our number system, each digit carries a place value or multiplier based on its position. For example, the number 205 has five ones, no tens, and two hundreds. Each position to the left carries a value ten times greater, because our system is based on the number 10. Got it? Same thing with **binary numbers,** but they are based on powers of two instead of ten. So the last digit in a binary number is the number of ones—just as in our system. But the next place to the left is the number of twos (not tens), and the next place is the number of fours, then the number of eights, and so on! If we use eight **bits** or places, the places have these values:

128s	64s	32s	16s	8s	4s	2s	1s

So how would we write our number 205 in binary? Start at the left and see if that place value fits in the number of choice. That is, subtract 128 from 205. Since the place value fits into the number, put a 1 in the first space.

1							

After we subtract 128 from 205, we have 77 left over. Now, 64 can fit in 77, leaving 13. But 32 can't fit in 13, so the next two places get a 1 and a zero.

1	1	0					

Keep going all the way down. Sixteen won't fit into 13, but 8 will with 5 left over. Four will fit into 5, but that leaves only one, and two will not fit there. So the final result is

1	1	0	0	1	1	0	1

Wow! So, 205 in our usual decimal notation is 11001101 in binary!

But why would you want to go through this!? It's simple—**data can be transferred in binary numbers with perfect precision!** All a computer or CD player or digital TV or digital phone has to do is detect ON or OFF, or YES and NO, or ONE and ZERO, over and over again. And that can be done much more easily than trying to tell the difference between a displacement of, say, 205 units and 206 or 204!

To make audio CDs, the analog signal is sampled 44,100 times each second, so the interval between the times in the picture on the other side is only 0.0000227 sec.

An audio CD or a DVD or Blu-Ray disc stores that series of ONs and OFFs as a series of reflective smoothspots and non-reflective pits or divots. A sharply focused laser beam shines onto the disc surface, and as the disc spins underneath it, the laser light reflects back to a detector, or it doesn't. A reflection from a smooth spot counts as an ON or a ONE, and of course the lack of reflection from a little pit is an OFF or a ZERO.

Recall that any television—high-definition or old-school analog—can only produce light that we call red, green, and blue. In a digital television, or a computer monitor, each color can have a range of 256 possibilities, from no color at all (zero) to full brightness of that color (255). That makes a total of $256 \times 256 \times 256$ or 16,777,216 color combinations!

Special combinations of ONEs and ZEROs on the disk tell the player when a new line begins for a video signal, when a new track or chapter begins, and other stuff that the player needs to organize that stream of ON and OFF pulses.

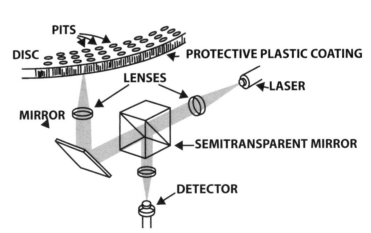

Computer hard drives work in the same way, except they produce and detect magnetic spots on a metallic surface. CD or DVD burners use a laser to actually burn or etch tiny pits into a thin reflective surface. Erasable or re-writeable optical disks, remarkably, allow the laser to melt and reform a shiny spot from a burned pit. They write more slowly, since they have to accomplish more.

These devices are only possible since the invention of laser diodes, tiny solid-state lasers on a single semiconductor chip. The first generation of videodisc players in the 1970s used expensive gas lasers, and a foot-wide disc was needed to record a single movie, as you see on the February, 1977 cover of *Popular Science*. The principle was just the same, though.

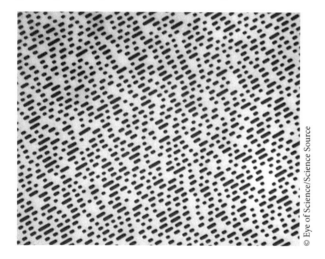

A microscopic view of the pits and smooth areas in an audio CD. Each pit is about 1/30th the width of a human hair. The pits are smaller and closer together on DVDs, and even smaller still on Bluray disks.

Graphing a Wave

Name _____

Section _____ Date _____

A vibration is a wiggle in time; a wave is a wiggle in space *and* time. In this exercise, you will explore that connection by transferring the motion of the second hand of a clock (a vibration, even if a slow one), to a wave.

Print out the image of a clock and open the spreadsheet called Grapher. You will find both of them on the course page in Learn.

Use a metric ruler to measure the displacement of the tip of the second hand every five seconds. That's the distance above or below the center line that connects the 3 and the 9. Count the number as positive if the tip is above the line (towards the 12) and negative if below (towards the six). Enter each value in the tinted boxes in the Grapher tool and it will immediately plot them for you.

Then answer the questions based on your graph.

1 | What is the period of this wave, that is, the time for one complete cycle, in seconds?

2 | What is the amplitude of the wave?

3 | If we had started from some other initial position—say, when the second hand was on the 3—would the period still be the same?

4 | Again, if we had started when the second hand was on the 3, would the amplitude still be the same?

5 | So—again, if we had started when the second hand was on the 3, what would be different?

6 | Using the period from Question 1, find the frequency of the wave. That frequency is of course way too low for us to hear. Please round you answer to three digits, not counting the starting zero; that is, round it to four decimal places total.

7 | If you had used the same picture of the same clock but had graphed the minute hand instead, what would change about the graph?

8 | If you had plotted the motion of the second hand on your watch instead of the clock that you printed out, what would change about the graph?

9 | Since you are plotting displacement on the y-axis and time on the x-axis, this is an example of a _____ graph.

10 | Are you old enough to remember the 1990s television series *Home Improvement*, starring Tim Allen? He was always saying "more power" and blowing up one electric tool or appliance after another by trying to make it work faster.

Suppose that he was able to make a clock run so fast that the second hand made one full lap in only one-tenth of a second (0.10 s), and suppose that it tapped a water surface each time it reached the bottom of its motion. If the waves it produced were 4.0 cm apart, at what speed (in cm/s) would they travel?

Adding Waves

Name _____

Section _____ Date _____

You might know from your own sad experience that two cars can't be in the same place at the same time. But two waves **can** be in the same place at the same time, and usually many waves exist in the same place at once. Think of how many light waves are striking your computer keyboard now from your screen and each of the light bulbs around you, or how many different sounds you might be hearing now. Or think about all of the radio stations in this area, and how all of their waves are flooding into your radio antenna at the same time.

Waves add through what physicists call the **Superposition Principle:** the displacement when two waves interact is just the algebraic sum of the individual displacements, keeping in mind the positive and negative signs on the displacements.

Try it for yourself! Open a new window on your computer and open the Excel spreadsheet called **Wavemaker** from the course page on Learn. With it you can choose the frequency, amplitude, and phase of as many as four waves and instantly add them together. The spreadsheet uses the Superposition Principle to combine the displacements of the waves at 120 time increments along the *x*-axis.

Just enter the frequency and amplitude values into the tinted boxes. As soon as you press Enter or move the cursor off the tinted box, the change takes effect. Note that if you enter **zero** for the amplitude of any wave, you effectively turn it off or eliminate it.

Please make the changes suggested below. I hope you will want to take some time to play around and ask yourself some "What if …" questions!

1 | Since time is on the *x*-axis, these are examples of what kind of graphs?

2 | Choose any frequency and amplitude within the ranges suggested in Wavemaker for Wave 1. Then change the phase to 90 degrees. Describe what happens to the wave.

3 | Keeping those same frequency and amplitude settings, now change the phase to 180 degrees. Describe how that new wave looks, compared with the original wave when the phase was zero.

4 | Reset the phase to zero, and set the frequency to 100 Hz. Reading from the graph, what is the period of the wave? (You could measure the time from when the wave first starts upwards to when it starts upwards again, or from one peak to the next.)

5 | Now add a harmonic to that wave. For wave 2, enter a whole number multiple of 100 Hz—that is, any whole number of hundreds of hertz. Set the amplitude smaller than you the value you used in Question 4. What is the new amplitude of the combined (bold, black) wave? (It will be different than the amplitude of the original single wave—but also different from the sum of the two amplitudes that you entered!)

6 | What happened to the side-to-side position of the peak of the new wave, compared with the original single wave?

7 | Here's a really important one—what is the period of the new, combined wave, from the first tall peak to the next one? (If your answer is **not** the same as Question 4, go back and check it again!)

Name _____

Section _____   Date _____

8 | Now leave the frequencies of both waves the same, but change the amplitude of the second wave to something larger than the amplitude of the first wave—say, two or three times as large. Now what do you find for the period of the new, combined wave? (I hope you are surprised by the answer!)

Add other harmonics by entering other multiples of 100 Hz for the frequencies of the third and fourth waves. No matter what frequencies or amplitudes you choose, the period of the combined wave should remain the same!

9 | Now for something completely different. Turn off waves 3 and 4 (set their amplitudes to zero) and enter a high frequency, such as 2000 Hz, for wave 1. Then enter something a little lower, say 1900 Hz, for the frequency of wave 2. Enter the same amplitude for both waves. What do you call the resulting pattern, where the intensity rises and falls as the waves drift in and out of step with each other?

10 | Increase the difference between the two frequencies. Leave one at 2000 Hz, but decrease the second one. What happens to the resulting pattern?

Exploring Resonance

Name _____

Section _____ Date _____

We had the good fortune of spending Spring Break in 2002 in Scotland, where my wife's brother, John Idoine, was studying while on sabbatical leave from his position as chairman of the Physics Department at Kenyon College.

In the beautiful city of Inverness the river Ness flows out from the famous Loch and into the North Sea. The river is spanned by several bridges, including this one, above, for pedestrians. We estimated that the central span between the towers was about 160 feet, or about 50 meters.

As a physics experiment most of us in the group went to the exact center of the span and started jumping up and down at a rate that matched the natural rhythm of the bridge. In this photo to the right you see my nieces Christine (left, in front of me) and Laura, with their brother Johnny and their dad, John.

My wife Linda took the picture with some difficulty, as the bridge was shaking quite a bit. My sister-in-law Debby was too embarrassed to be seen with us; she left the bridge and waited in a small park at one end, acting like she didn't know us.

Looking along the Iverness pedestrian bridge. From left to right: my niece Christine, me, my nephew Johnny, my brother-in-law John, and my niece Laura.

We created a transverse standing wave with nodes at the support towers and the antinode in the center, where we were standing.

I counted 24 full vertical oscillations in 15 seconds with an overall vertical motion of about six inches. (It was enough to make the towers sway!)

1 | Why was it essential that we jumped at that rate?

2 | What was the vertical amplitude, A, of the bridge, in inches?

3 | And how much is that amplitude from Question 2 in meters? (One inch is the same as 2.54 cm, and 100 cm make one meter.)

4 | Why was that quantity, the amplitude, especially difficult to measure accurately? (HINT: It's not because the bridge was moving so fast, or that the distance was so small. Think about it—where was I while measuring the amplitude?)

5 | From the information I gave you, what was the bridge's frequency of motion, f? (That is, how many vibrations or parts of vibrations did the bridge make in one second?)

6 | What was the period, T, of the motion, in seconds? (That is, how much time passed for each whole vibration of the bridge?)

7 | Knowing that the bridge was oscillating in its fundamental mode or first harmonic, use the information I gave you to estimate the wavelength in meters of the wave that we created in the bridge.

8 | Estimate the speed of the wave in m/s as it traveled down the bridge.

9 | Suppose the fabled Loch Ness Monster happened to swim by, grab the **center** of the bridge where we were standing, and shake it up and down at twice the frequency that you calculated in Question 5. That's the frequency of the second harmonic. Why would a standing wave **not** form on the bridge in that case?

10 | Following up on Question 9, what is the lowest frequency at which Nessie **would** have to shake the bridge at its center to make a standing wave other than the fundamental or first harmonic, that you calculated in Question 5?

Exploring Harmonics

Name _____

Section _____ Date _____

For each of the "frog-on-a-post" waves at top (A–C), match it with the Fourier analysis graph of that wave on the bottom (D–F) and with the source, at the left.

1 | Which two are of a tuning fork?

_____ & _____

2 | Which two are of an organ pipe?

_____ & _____

3 | Which two are of an office fan?

_____ & _____

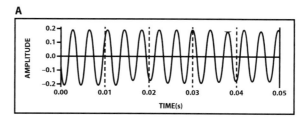

A

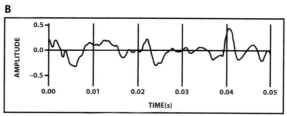

B

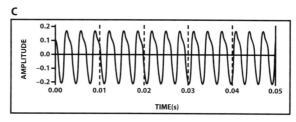

C

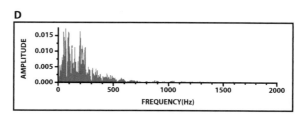

D

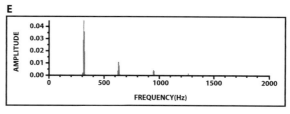

E

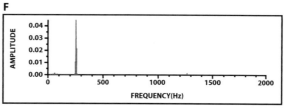

F

The graphs to the right show the frog-on-post graph and the Fourier analysis of me singing an "oooh" and an "eeeh."

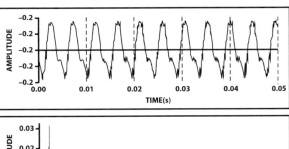

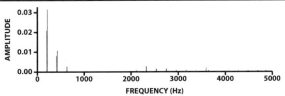

4 | Which pair is the "oooh?"

_____ the top pair

_____ the bottom pair

5 | How do you know which pair is the "oooh?"

6 | Which pair of graphs shows a sound of a higher pitch?

_____ the top pair

_____ the bottom pair

7 | How do you know which pair is the higher pitch?

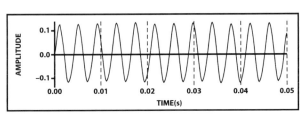

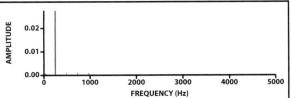

★ 8 | From the graphs on the right, explain why it's important that human hearing is most sensitive between about 2000 and 4000 Hz.

9 | The computer automatically scales the amplitude so that the waves seem to be the same intensity. But suppose the sound in the top pair registered 60 dB and that in the bottom pair measured 40 dB. How much more intense is the top pair?

_____ 10×   _____ 20×   _____ 50×   _____ 100×

10 | Using those same numbers, how much louder would the upper pair sound?

_____ 10×   _____ 20×   _____ 50×   _____ 100×

Applying the Speed of Light

Name _____

Section _____ Date _____

Do some Internet research to find these answers. You'll need a calculator for number 10.

1 | How did Galileo try to measure the speed of light?

2 | Approximately when did he conduct his experiments on the speed of light?

Galileo Galilei, 1564–1642, made the first known attempt to measure the speed of light.

3 | What did he conclude?

4 | How did Dutch astronomer Olaus Roemer measure the speed of light?

5 | When did he make his measurements?

Olaus Roemer, 1644–1710. His first name is sometimes written as Ole, and his last, as Rømer. What were they thinking with those wigs for me?

6 | And briefly, how did Armand Fizeau measure the speed of light in 1849?

7 | What is the best current value for the speed of light, in m/s?

French physicist Armand Fizeau, 1918–1896.

8 | What does our precise knowledge of the speed of light and radio waves have to do with the GPS (the Global Positioning System).

9 | Why did several teams of Apollo astronauts leave glass reflectors, called retroreflectors, on the moon's surface? (This is, by the way, one of the best proofs that our astronauts actually went to the moon, which some people argue never happened.)

10 | Dr. Ceri Brenner, a physicist from the United Kingdom, is one of many scientists working on power from nuclear fusion. At the National Ignition Facility in California, they blast tiny pellets of frozen hydrogen from all sides with 192 simultaneous laser beams. Each laser pulse lasts for only 20 nanoseconds. That's 2.0×10^{-8} s, or 0.000000020 s. How many meters can light travel in that time? (Round your answer off to two digits, please.)

Mixing Colors

Name _____

Section _____ Date _____

Adding Colored Light Beams

Helpful Hints

Additive primary colors are all one syllable. As **lights** they add to make white. As **pigments** they absorb light of both other primary colors. (For example, red pigment absorbs both green and blue light.)

Subtractive primary colors are all more than one syllable and have a single purpose: magenta absorbs green light, yellow absorbs blue light, and cyan absorbs red light. All three together absorb all light and look black.

Suppose beams of light containing the frequencies that we see as the indicated colors shine together on a white surface. What color will you see?

1 | red + green looks   _____

2 | green + blue looks   _____

3 | magenta + green looks   _____

4 | red + cyan looks   _____

Mixing Pigments

Read the hints above, and then imagine that white light is shining on a mixture of these pigments or dyes. What light will be reflected, if any?

HINT: These are pigments, not lights.

5 | yellow + magenta reflects   _____

6 | magenta + cyan reflects   _____

7 | yellow + cyan reflects   _____

8 | magenta + cyan + yellow reflects _____

9 | red + green reflects   _____

Color Printing

What subtractive primary colors must be mixed to produce the appearance of each of these colors?

10 | red _____

11 | green _____

12 | blue _____

13 | white _____

14 | black _____

Mixing Lights and Pigments

Read the hints on the first page and imagine that light of the color indicated in the left column is shining on the pigments in the right column. What color will you see?

Incoming Light		Shining on Pigments	
15	blue light	→	cyan + magenta pigments looks _____
16	yellow light	→	cyan pigment looks _____
17	red light	→	magenta + yellow pigments looks _____
18	green light	→	yellow + magenta pigments looks _____
19	magenta light	→	red pigment looks _____
20	cyan light	→	magenta pigment looks _____

Applying Reflection and Refraction

Name _____

Section _____ Date _____

Choose the best answer for these questions.

1 | What size flat mirror, mounted on a wall at eye level, do you need to see your whole self?

2 | This sign in the restroom of the 12 Bones barbecue restaurant in Asheville, North Carolina says "See— good barbecue makes you skinny!" How can the mirror make me look skinny but of normal height?

3 | Is this photo taken looking into the front (top) side of a spoon, or the back (bottom) side?

4 | She is standing very close to this fun-house mirror that appears to be concave from top to bottom, but normal from side to side. What can you say about her image?

5| Here is a sketch of a light ray passing through a glass prism. Since the ray passes straight through without bending, how does the index of refraction of the glass (n_1) compare with the index of refraction of the medium (n_2) surrounding it?

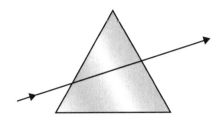

These sketches show light rays as black arrows refracting when moving from one medium to another, where the thick black line is the surface and the dotted line is the perpendicular or normal line. Some are drawn correctly and some are not. For each one, check the best answer.

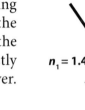

(6)

6| The ray in number 6

a| is correct as drawn.

b| should bend to the left.

c| should bend to the right.

d| should go straight without bending.

(7)

7| The ray in number 7

a| is correct as drawn.

b| should bend to the left.

c| should bend to the right.

d| should go straight without bending.

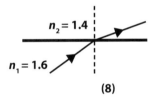

(8)

8| The ray in number 8

a| is correct as drawn.

b| should bend to the left.

c| should bend to the right.

d| should go straight without bending.

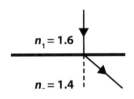

9| The ray in number 9

a| is correct as drawn.

b| should bend to the left.

c| should bend to the right.

d| should go straight without bending.

10| In which of the numbered cases above would light reflect completely from the surface if the angle of incidence grew larger and larger? Choose all that apply.

a| 6   c| 8

b| 7   d| 9

Working with Lenses

Name _____

Section _____ Date _____

Choose the best answer for each question. Use this drawing for the first three questions.

1 | This drawing represents case _____ on the Lens Chart.

2 | The lens is a

    a | diverging lense.

    b | converging lens.

3 | How does distance from the letters on the page to the lens compare with the focal length of the lens? The distance from the letters to the lens is

    a | greater than the focal length.

    b | equal to the focal length.

    c | less than the focal length.

4 | In a movie projector in your local cineplex, the frames of film go into the projector

    a | head first or upside down.

    b | feet first or right side up.

5 | Again in a movie projector, where is the film positioned?

    a | A little under one focal length behind the lens.

    b | Exactly one focal length behind the lens.

    c | A little over one focal length behind the lens.

6 | In a traveling or following spotlight in a musical theater production, the brilliant xenon lamp positioned

    a | a little under one focal length behind the lens.

    b | exactly one focal length behind the lens.

    c | a little over one focal length behind the lens.

7 | As a tourist you are pretending to take a photo of a splendid castle in the distance, but you really want to capture the very attractive person not far from you. If you have already focused on the castle, but now you want to focus on the cute person, you would need to

a | leave the lens alone.

b | move the lens out farther from the camera.

c | move the lens in closer to the camera.

8 | Suppose the distance from the candle to the lens is 40 cm, and the distance from the lens to the screen is also 40 cm. What is the focal length of the lens?

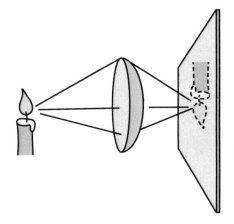

9 | My older son Ben is quite a good juggler. He got in some trouble once when he left a set of clear acrylic juggling balls, like this one, in the back end of our Ford Escort station wagon while it was parked for a long time in the sun. Why do you suppose I got a little mad at him?

10 | The flowers in the photo stand out in sharp focus, but the background is a soft blur. That means that the camera lens that took the photo was set for a

a | small aperture and a low f/ratio.

b | small aperture and a high f/ratio.

c | large aperture and a low f/ratio.

d | large aperture and a high f/ratio.